D0896937

CONTENTS

A curious cat inspects the depths of a privy in Puddletown.

FOREWORD

It was as if some forgotten nerve in the county's collective psyche had been triggered when local newspapers publicised my search for memories of Dorset privies. A trickle of letters became a flood, bringing confirmation that people remember their back garden outhouses with an affection that the antiseptic modern bathroom will never attract. Clearly, the privy has a definite yet hitherto unsung place in Dorset's heritage, which surely makes it important to preserve at least some of the few remaining specimens. It is sad to note the readiness with which even the most picturesque ones are demolished, not always by newcomers to the county. As one concerned lady told me, 'The damage being done is desperate.' At least I have managed to record some of them, and the wonderful tales people tell, before they are lost for ever.

I am indebted to all the kind folk who contributed their stories, allowed me to take photographs or resisted the temptation to call the constabulary when I peered over their garden fences with the outrageous excuse that I was looking for old privies. Now do you believe me? My thanks also go to Dorset newspapers and local history and genealogical societies for their invaluable assistance. The pictures of an earth closet and of the Reverend and Mrs Henry Moule are reproduced by kind permission of the Dorset Natural History and Archaeological Society at the Dorset County Museum. And an especial word of thanks to my wife Sandra, who encouraged me to tackle this project, then patiently supported me to its conclusion.

Apologies are offered to those who insist the proper Dorset term was not privy but mixen; my excuse is that an equal number of people say a mixen was the compost heap where a privy's productions were mixed with other organic materials. I would rather not take sides.

IAN FOX

5

Late autumn sun illuminates this little beauty in the back garden of a Bagber cottage. I particularly admire the roof tiles.

[1]

A WEE LOOK BACK

I met a man in Abbotsbury who remembers the days before mains sewerage reached the village. We stood in his back garden, halfway up the path, and I admired his sturdy stone privy and a big old bucket, both of which had given faithful service before the coming of the WC made them redundant. Cool and dark, the privy now provides an ideal store for his potatoes but still he refers to it fondly as The Old Shitery.

His preferred method of disposal was to dig the bucket's noisome contents into his vegetable plot at the top of the garden. Not everyone did the same.

'Lots of folk just tipped it in the stream, same as they did in Portesham and plenty of other places,' he said. 'It went downriver and finished up in the Swannery. Or else they'd watch no one was looking and then nip out and empty the bucket down the storm drain in the street. That was all right if we had rain to flush it into the stream but in dry weather it just lay down there and piled up. Them drains used to stink something rotten in summer.'

Those good people of Abbotsbury faced an age-old problem. The question of how to get rid of our most basic product has perplexed the inhabitants of these islands for centuries. We boast of having given the world the WC but generally our record has been pretty abysmal. Not until Victoria's reign, when overcrowding and appalling sanitation resulted in some nasty outbreaks of cholera, did we begin to take things seriously. And it is a sobering thought that even in the second half of the 20th century, thousands of Britons still had to do their business in an outdoors privy, using a bucket or a hole in the ground.

Their forebears would have squatted behind the nearest bush

Sited halfway up a garden in Abbotsbury, and finding a new lease of life as a potato store, 'The Old Shitery' formerly gave long and loyal service as the family privy. 'It was damned cold coming up here in the middle of the night,' remembers the owner.

with a handful of grass or moss, or perhaps obeyed the order to retire 'a bow's shot away' from any habitation. As settlements grew it made sense to dump everything, directly or indirectly, into communal latrine pits, which incidentally have since provided happy hunting grounds for archaeologists – they love sifting through ancient ordure to discover what our ancestors ate.

Castles and the homes of the wealthy had their 'garderobes', small rooms or alcoves that sometimes were built into the thick stone walls, preferably near a fireplace to afford some relief from the cold. Clothes might be stored here but more important was the seat-hole, beneath which was nothing but the open air. Exposing your bare bum on this draughty aperture, high on the castle wall, must have been torture on an icy winter's day. However, it did allow matters to be expelled instantly from the build-

The millstream still flows beneath the remains of a 13th-century garderobe tower alongside the Norman House at Christchurch Castle. This was much nicer than an external garderobe; matters were hidden as they plummeted into the water, emerging through the arch to bob away downstream to the sea.

ing, often falling a considerable distance to join an extremely unpleasant mound on the ground, or perhaps splashing into the moat, a stream, a defensive ditch or a huge barrel. Some lowly servant or grateful farmer would gather it up from time to time.

Medieval monks strove to avoid disease within their confined communities and often devised means of flushing away their waste, building the 'necessarium' or privy block over a river or a specially diverted watercourse that ran beneath the seats before rejoining the main stream. Some monasteries installed ingenious methods of using stored rainwater to flush the necessarium gully. Others simply had latrine pits and in these have been found torn rags that passed for toilet paper, together with abundant seeds of buckthorn berries, a common purgative.

As the Middle Ages progressed and towns became more crowded, prodigious amounts of sewage had to be shifted. The experts at this unenviable skill were the 'gong-fermers', whose name derived from 'gong', a privy, and 'fey', meaning to cleanse. They could command high fees but earned every penny. Their job usually entailed removing the filth to massive pits, to the river or into the countryside by the barrel or wagon load. Sometimes, though, they might be standing waist-deep to clear a brimming cesspit or blocked sewer beneath a communal privy.

The towns stank, thanks partly to the habit of emptying chamber pots directly into the street out of any convenient window or door, a charming practice that continued for hundreds of years, despite the law's efforts to prevent it. Dorset court records contain many references to prosecutions of people who accumulated 'noisome dung mixons' outside their front doors or 'cast out their Filthy Excrements to the Annoyance of Neighbours'. Some towns appointed 'scavengers' to literally scrape a living by removing at least some of the mess from the streets.

Sir John Harington, godson to Queen Elizabeth, is credited with inventing the first water closet with moving parts. At least two examples of his crude apparatus saw the light of day in 1596 but it failed to catch on. Not until the 18th century, when piped water supplies became more prevalent, was any real progress made. Alexander Cummings was granted the first patent for a valve water closet, in 1775, and three years later Joseph Bramah produced a superior WC which remained unsurpassed for more than a century. We still use his surname when praising anything of the best quality.

Those early models had cast iron bowls with wooden surrounds, rather like a flushable commode. WCs that we would recognise, made from porcelain and with high cisterns, began to appear in the late-19th century. This was the Golden Age, when such manufacturers as George Jennings, Thomas

This Hogarth illustration of 1738 shows the hazards of walking beneath an open window when the chamber pots were being emptied!

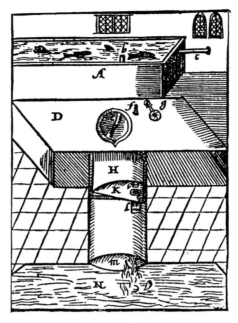

Sir John Harington's design for a water-flushed privy, from his *Metamorphosis of Ajax* (1596). The fishes in the cistern (A) are drawn for fun. 'D' is a board with the seat hole. Water flushed through the pot (H), down the sluice and into the bottom vault, which Harington recommended should be emptied at noon and at night.

Twyford, John Shanks and Sir Henry Doulton competed to produce efficient flushing systems and the most attractive bowls, highly ornamented and often in fantastic shapes. The eponymous Thomas Crapper made his mark at this time, not by inventing the WC as many believe but by perfecting his Valveless Water-Waste Preventer, a clever device designed to eject a set amount of water at each flush. It forms the basis for most modern WC systems.

Many of the early WCs simply discharged untreated sewage into nearby rivers. Others emptied into crude cesspits alongside

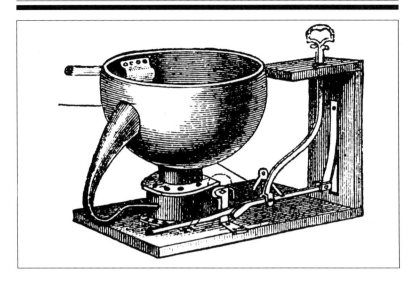

Joseph Bramah's water closet of 1778. The handle operated an emptying valve beneath the iron bowl. Intended to be encased in wood, this WC set the standard for more than 100 years.

homes or even beneath their floorboards, from whence they stank abominably. The majority of households still had only pots or the traditional privy, usually just a pail or a pit (a 'vault') beneath a wooden seat. But providing piped water and proper sewerage systems to serve every household was expensive, particularly in rural counties such as Dorset. Bitter local opposition was mounted when ratepayers realised what such refinements would cost, while many people genuinely believed their privies to be the more hygienic option.

The situation in Bridport was not untypical. Controversy raged for years, even though a government inspector in 1864 criticised 'the mass of putrid matter that lies in its privy vaults and poisons the air by its decomposition'. A satirical broadsheet visualised the complacent 'Men of Bridport' singing a song which began:

I likes the smell of dirt,
It seems to do me good;
It raises up my spirits
And relishes my food.

A dung-hill at my door
And a cess-pool close at hand
Is the healthiest things in nature,
The sweetest in the land.

Nevertheless, sufficient men of Bridport decided they would rather not pay a shilling rate to improve matters and after a further eight years of dispute it took State intervention before the town got its drains.

Blandford had to wait even longer. Throughout the 1920s, the borough's Medical Officer of Health repeatedly denounced what he termed 'the old fashioned Pail Closet Conservancy System', saying it was Blandford's 'most serious sanitary defect'. Once a week, early in the morning – often at breakfast time – a contractor's horse-drawn carts visited every home to remove filled privy buckets and leave empty ones. Many houses were terraced, which meant the sometimes brimming pail had to be carried straight through, with results that can be imagined. Then the carts slopped away, bound for the dumping ground at Coward's Farm (now part of the Damory Down estate), followed by the council's water cart to sluice away the considerable spillage. Some elderly residents remember disinfectant also being sprayed on the roads. The council's own tanker lorries took over the collection in 1926, resulting in 'much improvement in the speed and cleanliness of removal'.

Blandford finally got a proper sewerage system in 1930, thanks to a government scheme that provided 75 per cent of

This 'sentry box' of corrugated iron at Long Crichel was in use until mains services arrived in 1956. Before then, all water was drawn in buckets from a well on the far side of the road. The privy pail was emptied in the garden. Alongside are the former wash-house and bake oven.

the cost. The council still had to borrow £45,000, a massive sum which took 44 years to pay off. Small wonder, then, that thousands of Dorset people elsewhere were doomed to persevere with the privy and its antiquated 'bucket and chuck it' system, sometimes into the 1960s.

As we shall see, they experienced mud, mishaps, midges and some memorable moments of merriment on their trips down the garden path.

[2]

Country Ways

I would like you to meet two ladies who remember precisely what it was like to grow up with a privy in the back yard. Their childhood was spent in rural Dorset during the early years of the 20th century, when an outhouse in the garden was accepted as part of a way of living that nowadays seems unthinkable.

Here is what Mrs Kitty Sinnott told me. 'I was brought up in the tiny village of St Giles, well over 80 years ago. It's Wimborne St Giles now. We called our privy "WC" or simple "double-U". Years later I learned that WC stood for water closet but ours had no water, just a smelly old bucket under a wooden seat with a hole.

'My grandmother called hers the Houses of Parliament. She had a calendar and pictures on her wall. Ours was situated well up the garden. I always left things until the last minute and in the mornings you'd find me running up the small slope in a great hurry, sometimes arriving only just in time. I know where jogging started!

'There was no toilet paper as there is today, just squares of newspaper on a string. I remember my mother saying, "Don't use so much paper – it fills the bucket." She had to empty it frequently into what used to be a saw pit belonging to my father's business (he was a wheelwright, carpenter and undertaker), although the saw pit wasn't used as such in my day. Father's workmen had their own privy. It was a two-seater over a pit. We were not allowed in, but I had a peep.

'My dear old Aunt Flo also had a two-seater over a pit, which was emptied when necessary from the back. One day, there was Aunt Flo about her morning exercise, not knowing the man had come to do the emptying. He saw a bare backside and couldn't

16

One can only guess at the age of this venerable two-holer. Low and cramped (the camera is literally at the front door), it is tucked into the garden wall of a 16th-century cottage at Bishops Caundle. Using the toilet was often a social occasion, judging by the number of two- and three-holers, while six-holers are not unknown.

resist giving it a poke. She was so embarrassed and said, "I'll give you half a crown not to tell anyone about it."

'My young sister was scared to go to the double-U at night, so my other sister and I took it in turns to go with her. Once there, she would sit and sit, and I would be outside, shouting, "Hurry up, it's cold out here." All was so quiet up there. Sometimes you'd hear a crunching noise – a cow would have pushed through the hedge and was eating the cabbages.

'Ah! The shelf in the privy – or was it the top of a beam? That was a grand place for hiding a packet of Woodbine cigarettes. A smoke was always a help in there.'

Mrs Sybil Dougan was 77 years old when we met. Her memories

of childhood were also as clear as if it were yesterday. One of seven children, Sybil was brought up in the village of Holden-hurst, on the outskirts of Bournemouth, in the days when her father paid 10 shillings a week rent for their home.

This is her story. 'We had a wooden privy with a bucket and a wooden seat, right at the very top of the garden, up a long narrow path. It was a long walk up there. You knew that if you wanted to use the lav you had to get going in good time, especially at night, because it was fatal to wait too long in case you never made it. We used to call it the lav, but my mother didn't like that. She used to say, "Call it the privy – it sounds much better."

'Mother used to scrub the inside of the lav every week – the floor, all the wooden walls, everything. She was a right fussy woman and very particular about the seat. After scrubbing the lav, she'd tell us children, "Now make sure you don't get it dirty. Don't wet on the seat or anything else – it's clean as an ant's tooth now." I never knew how clean an ant's tooth was but that old seat was pure white where she scrubbed it so much.

'Every night my father would empty the bucket. He would go and dig a very deep hole in the garden and bury it, and oh! he grew such beautiful vegetables. There used to be a fête at Hurn Court, where Lord Malmesbury lived, and Dad used to take first prize for his vegetables – celery, carrots, onions, marrows, everything he took there, and he said it was all down to the manure he put on the garden.

'Unfortunately for my father, though, he didn't have a very strong stomach and couldn't stand the smell, so whenever he emptied the bucket he used to wear a gas mask, which we all thought was really comical.

'A rat lived in our privy – it actually lived up there, behind the toilet seat. It was there as long as I can remember. This great big rat used to sit in the corner and look at me while I was on the privy. I was really, really frightened of it. If we asked to go to the privy at night, Dad used to say, "You know where it is, go on up there." We'd say, "But Dad, there's a rat in there," and he'd say, "Well, that's its home, isn't it. It won't hurt you." He was determined we wouldn't be frightened of rats, or anything else for that matter. Eventually we got into the habit of saying "Hello, rat" when we went into the lav, but I was always scared of it.

'Dad wouldn't even let us use a torch to get up the garden at night. I remember him saying, "You've been up there enough times to know where you're going, wench. You can feel your way up there." And so we used to go all the way up the garden in the pitch black night, you couldn't see your hand in front of your face sometimes, and there was no light up there at all, not even a candle. I remember sitting there and seeing the rat's eyes shining at me in the dark.

'We had to go in the middle of the night and in all weathers. Dad wouldn't let us have a pot under the bed. He used to say it made you lazy. His law was: "If you don't like using the privy at night, don't drink anything before you go to bed."

'There was a lift-up latch but once you were in you couldn't lock the door because it didn't have a lock. You just went up there and took a chance while you were inside.'

Sybil also remembered, without much affection, the facilities at her little school in Holdenhurst. 'We had privies there, too. There were only about 60 or 70 pupils. It had eight outside toilet cubicles with a bucket in each one. The girls had to share them with the boys, there were no separate ones for us. There were wooden seats over the buckets but no lids to keep the smell in, so it was terrible out there. Also, some of the boys came from very dirty families and weren't too fussy about things. I was lucky, we didn't live very far away, so I used to bottle it up until I got home rather than use the school ones. If any of us children were naughty in class, the teacher used to send us out to stay in the toilets as a punishment.

19

When Colin and Marilyn Beale bought an 18th-century cottage in Wareham, their solicitor added a clause stating that there was 'a place of quiet contemplation at the bottom of the garden'. To the right of the lamp-post, almost hidden amidst foliage, the tiny privy actually stands within their neighbour's boundary wall.

'Every night at 5 o'clock an old chap with a horse and cart used to come round to empty the buckets. He had some old milk churns in the back of his cart. He used to go into the toilets, lift out the full buckets and bring them out of the school and pour them into the milk churns.

'There were no lids on the churns, so you can imagine the smell. It was a right horrible pong. If the wind was the wrong way they could smell it right across at Throop Mill. Our house was quite near the school so we used to get the full belt of it. I remember calling to my Mum once, "Go inside and hold your nose, here comes the shit cart." She gave me a right telling-off for using such language.

Part privy, part pigsty – a common Dorset practice. The animals had one half of the brick building, humans the other, as here at Lytchett Matravers.

The door is missing but here at Tolpuddle is a fine example of a combined pigsty and privy. The actual privy is hardly wider than the door – pigs occupied the remainder of the brick building under the pitched roof.

21

'The school used newspaper as toilet paper and two or three times a week a number of children had to sit in the classroom, cut the paper into squares and thread it onto string.

'It was the same at home. Once a week, for about an hour, my father used to get all us children sat round the table and we had to cut up the squares of newspaper, about the size of today's toilet paper, and put it on string. Then it was hung on a nail in the lav. It was quite a performance but we got enjoyment out of it.

'Later on we moved to Throop and lived there for four and a half years. It was like moving from the ridiculous to the sublime. Didn't we have lovely toilets there! Two of them, both indoors and with the flush, too. We thought we were in Heaven.'

[3]

BACK TO NATURE

It is one thing to casually flush your loo and despatch the contents away round the S-bend, out of sight and out of mind, but it was a whole different story in those pre-flush days. What you had done was what you were left with and the responsibility for disposing of it was all yours. Some folk used the services of a weekly bucket collection, either by the council's 'honeycart' or through a private contractor. This could be a mixed blessing. The good news was that someone else had the job of getting rid of the stuff. On the other hand, a week could be a very long time to wait, especially if the family was large and productive and matters were piling up.

The situation became particularly fraught in town houses with little or no spare garden in which to dispose of the surplus, so it became second nature to keep one eye on the level in the bucket while anxiously awaiting the appointed day for your saviour to arrive with his cart – rather like a beleaguered garrison watching for the relieving cavalry to appear over the hill.

The most common (I hesitate to use the word popular) method of disposal was burial in the garden. It cost nothing, was relatively simple and had the added benefit of enriching the good earth, which rewarded your labours with bountiful vegetables and flowers. Dorset folk recycled their waste with a dedication that nowadays would have Friends of the Earth weeping tears of joy.

Evidence of the benefits of the old 'bucket and burial' system comes from Mr Rawles of Wareham, who remembers helping his gran when he was a lad. 'I used to have to empty the bucket when it was nearly full. I say "nearly" because if you were late doing it you got into an awful state, if you see what I mean. I had

23

Mrs Irene Webb outside the old family privy at Winterborne Stickland which her mother used until her death in 1994, aged 93. 'After the bucket had been emptied in the garden, the spaniel would dig the stuff up and roll in it, so we had to cover it with a sheet of galvanised iron, weighted down with logs and big stones.'

Privies in blocks of three were still serving terraced cottages at Durweston in the 1960s, when the weekly collection lorry charged half a crown per bucket. Most residents preferred to dig it into their gardens, where it could do some good.

to empty it into a large hole that my uncle dug in the garden, with garden waste and soil, which was called the "mixen", and when it was full he would plant marrow seeds on it, and you should have seen his marrows – they were whoppers.

'My Uncle Bill that lived just up the road used to do the same and one year he had to have a wheelbarrow to move his marrow. It was in the *Guinness Book of Records* for a while, and that is what I would call organic gardening.

'I remember my uncle came indoors one day and said to my gran, "What do you think of Mr Marshallsay, he's emptying his bucket in daylight." Gran said, "Well, so do we", to which he replied, "Yes, but theirs do stink." Those were the days.'

Some keen gardeners were grateful for any extra they could lay their hands on. Freda Lush of Dorchester remembers her schooldays in the 1930s, when she lived in the schoolhouse at Tincleton. Her mother was the headmistress. 'We had our own family privy in the back yard, with a wooden seat and an ordinary large bucket below. Over the wall were similar ones for the school children. Every Friday night, quite late, an old man from higher up the village used to come round with a large container on a hand cart and empty them. His garden, of course, was the best in the village. As far as I know, this was still going on until at least the end of the last war.'

While many people preferred to use their bucket to supplement the compost heap, bean trench or marrow mound, others simply wished to bury the stuff as quickly as possible. The interesting challenge was to find a spare piece of ground in the garden, or to remember where the last offerings had been laid to rest and so avoid the trauma of uncovering something best left undisturbed. I have heard of people marking their most recent burial plot with an upturned flowerpot, a garden gnome or a plastic sunflower. One man used a whitewashed flat stone, like a tombstone, bearing the warning 'Yer Tiz'.

Stout buckets were the mainstay of many privies. John Enderby (who, coincidentally, is an amateur archaeologist!) unearthed this one from his garden in Fontmell Magna. 'The night soil was probably dumped on my vegetable garden and accounts for its wonderful fruitfulness,' he says.

Mrs Kirk from Wallisdown was one of six children living in a cottage in Throop, where the privy was on the far side of the garden. 'My mother had to empty it every week by digging holes all over the garden. Using the privy was rather creepy at night and one of us would stand at the door with a torch to shine across the garden so the one who wanted to use the toilet could see the way and run as quickly as possible.'

Kathleen Mould recalls that when her father was selling his cottage, he answered the official query 'What sanitation?' with an equally formal reply: 'Private bucket at the bottom of the garden'. He served in the Royal Navy, so was away from home for long periods. 'That was when I, his daughter of seven to ten years, often found myself digging the hole in the garden to bury the contents of the bucket,' she says.

Emptying the bucket was not the most popular chore and the arrival of visitors was a mixed blessing. It meant additional work, as J. Symes reminds us. 'My sister married a naval lad from Burton Bradstock near Bridport in 1946. Their first home was a very small cottage in Donkey Lane which had an earth closet at the bottom of the garden. His garden produce was always most bountiful. Although they always made us welcome for holidays, I am sure my brother-in-law gave a sigh of relief when we went home. In the early 1950s they moved to a modern council house and a few years after that the council arranged for all the houses to be connected to the main sewers. I remember my sister saying that quite a few people were not at all keen to change over to the new system.'

With a little imagination it was possible to turn an otherwise mundane chore into something of an art form. Maureen Linnell remembers the challenge that faced her relatives in their remote farmhouse near Sherborne when, as recently as the 1960s, the landlady stubbornly refused to modernise the property. There was no electricity, no mains water and no drainage – which meant they had a Privy Problem.

'Being resourceful, they came up with an ingenious solution,' Maureen recalls. 'Their garden was large, well-drained and sloping. A number of paths were laid out which radiated from the back door, out across the garden like the spokes of a cart-wheel. The privy, which was made of wood and portable, looked like a cross between a sentry box and a sedan chair. Initially, it would be positioned at the end of one of the paths and it was moved on average once a week, so in the course of time it zigzagged from left to right across the garden until the length of each path had been used up.

'Before each weekly uprooting a small pit would be dug at the new destination, ready to receive the privy. The excavated earth

Although a river runs conveniently close by, I am assured that 'bucket and burial' was the order of the day with this sturdy specimen at Wootton Fitzpaine.

29

would be used to fill in what remained of the pit on the spot just vacated. Fine crops of vegetables grew on the triangular plot of ground between each path.

'The arrangement had two disadvantages. Would-be users tended to get disorientated, particularly on dark nights, working out which path was the current one to follow. More seriously, in stormy weather the portable privy, being lightweight, had a tendency to take off or to blow over – not much fun if it was occupied at the time, as the seat was an integral part.

'Eventually, the family built themselves a bungalow with all mod cons. They named it "At Last". Needless to say, it had no privy.'

[4]

THE CLERGYMAN'S CLOSET

Few people know that a Dorset vicar became famous for improving the humble privy. A statue of his friend, Thomas Hardy, occupies a prime position in the county town, but mention the Reverend Henry Moule to anyone in Dorset, or elsewhere for that matter, and the chances are they've never heard of him. And yet his name was once a household word, like Hoover or Biro.

Henry Moule was a determined evangelist who fought ignorance, vice, filth and disease on behalf of his parishioners. He battled with landlords, local authorities and even royalty to improve the lot of his flock. He was clergyman, teacher, inventor, engineer and entrepreneur, and he occupies an important place in privy history for perfecting and promoting the earth closet – a WC without water.

Moule was born in 1801 and became vicar of Fordington, a parish of some 2,000 souls on the outskirts of Dorchester, in 1829. He at once set about correcting the mess left by his predecessor, who had carried out only the most unavoidable duties. St George's church had fallen into disrepair, women had been paid to take Holy Communion and, as the parish clerk pointed out when Moule asked for water to perform a baptism, 'Last parson never used no water. He spit into his hand.'

The new parson was having none of this and proceeded to shake up the dissolute population, to their bewilderment and annoyance. He began preaching two Sunday sermons ('I never gave greater offence in the town,' he said later), castigated his parishioners' moral laxity from the pulpit, changed the church music and upset the choir so much that most of them resigned. His reforming zeal attracted a deal of hostility, particularly

Henry Moule was about 70 years old and famous for inventing his earth closet when this rather faded photograph was taken.

Mary Ann Moule, pictured here in 1869, was praised for her selfless efforts during the cholera outbreaks in Fordington.

when he persuaded a wealthy and influential woman to with-
draw her patronage from the 'sinful' Dorchester Races, so
ending an enormously popular social occasion. Rich and poor
alike were outraged to have their fun stopped by the meddling
cleric.

Moule's tenure at Fordington was one of tireless activity.
Besides improving St George's church, he built a Sunday
school, a day school and a new church, Christchurch in West
Fordington, putting much of his own money into the projects.
He experimented with animal husbandry, agriculture and ferti-
lisers, wrote hymns, books and innumerable pamphlets, made
several inventions, lectured on natural history, archaeology
and antiquities, helped to found the Dorset Museum and an
adult education institute, and involved himself in many cam-
paigns on behalf of the less-privileged.

Somehow, he and his wife Mary also found time to raise seven
children and, for 30 years, to accommodate up to ten youngsters
at a time as boarders at the Vicarage, where they were tutored
by Moule and his sons.

The events which had the greatest influence on his life
occurred in the middle of the 19th century, when many of Fordington's inhabitants existed amid squalor and social depriva-
tion. Their already atrocious living conditions had worsened as
displaced agricultural labourers, immigrants and travellers
arrived to seek work or shelter. It was a time of great social
upheaval and the authorities were overwhelmed by an influx
for which they were not prepared.

The population had risen to almost 3,150. They packed into
tiny cottages, tenements and jerry-built hovels, many of which,
said Moule, were 'utterly destitute of the ordinary conveniences
of life'. Often their only source of water for bathing, laundry,
cooking and drinking was the river Frome or the mill pond.
Into those same waters went their waste, either directly or indirectly. Sewage outfalls added the filth of Dorchester, mixing with

discharges from light industries and the many laundry businesses of the parish. In the absence of sufficient privies, people had no alternative but to use open street-gullies or the marshy fields. Privy buckets and slop pails were emptied in any place available, and stinking, stagnant pools lay all around.

Conditions were so bad that in later years, when writing *The Mayor of Casterbridge* and *A Changed Man*, Thomas Hardy chose Mill Street in Fordington (he called it 'Durnover') as his inspiration for 'Mixen Lane', with its squalor, vice and drunkenness.

Into this wretched parish came cholera. It struck first in 1849. In October of that year, Moule recorded that the average annual death rate of 62 had already almost doubled, to 120, through smallpox, typhus, scarlet fever and cholera.

Five years later, in September 1854, when 'King Cholera' again visited Fordington, Moule angrily condemned the deplorable conditions in the mill pond area, where about 1,100 people were crammed 'within a space that can scarcely exceed five acres'. He described how 'their filth is cast either into the open and wretched drain in the street, or into the mill pond, from which the people draw most of their water for washing and sometimes even for culinary purposes'. There were hardly any privies – incredibly, he knew of one that was shared by 90 people, another by as many as 100. 'In one square of 13 miserable cottages, there cannot be said to be any "convenience",' he added.

Moule also identified what he believed had caused the latest epidemic. When cholera swept through London's Millbank Penitentiary, 700 convicts had been moved temporarily to the barracks at Dorchester. They brought death with them. Disastrously for Fordington, the prisoners' clothing was sent there to be cleaned and two washerwomen had crammed the filthy garments into their tiny cottages.

First a child died and then a further 23 people of the 93 who contracted cholera. As terror gripped the inhabitants, Henry

Moule and his wife attracted praise for their selfless endeavours to prevent the disease spreading to Dorchester and to help the victims. It was while visiting one of them that the vicar saw a sight he never forgot. Heavy rain was falling and, as he knelt by the dying man's bedside, Moule was horrified when 'the overflow of the one privy for 13 families passed between myself and the bed'. Then more raw sewage erupted from the earth beneath the fireplace.

Such awful experiences during Fordington's two cholera outbreaks, coupled with his anger at the overcrowding, squalor and poverty, stung Moule into action. Most of the parish was owned – and neglected – by The Duchy of Cornwall and so it was to the President of the Duchy, Prince Albert, that Moule addressed a series of letters demanding drastic reform of the housing situation. He also ranted at local politicians in attempts to resolve the question of sewerage, a matter to which he now applied his inventive mind.

Impressed by experiments that showed how solids were absorbed and deodorised when animal manure was passed through earth, Moule applied the principle to human waste disposal. He invented an earth closet which he patented in 1860, proclaiming it to be superior to the water closets that were slowly being introduced.

What seems to have driven him was his belief that 'in God's providence there is no waste'; human manure was an asset that should not be flushed away but recycled as Nature surely intended. Mix it with dry earth to render it inoffensive, he urged. Householders could dig the resulting product into their gardens, while establishments that produced large amounts – boarding schools, for example – could bag it up and sell it by the ton as fertiliser. Town and city councils, he enthused, could reduce the rates if they forgot sewers and WCs, installed earth closets and collected the proceeds for sale. Moule himself charged £3 a ton to raise funds for local schools.

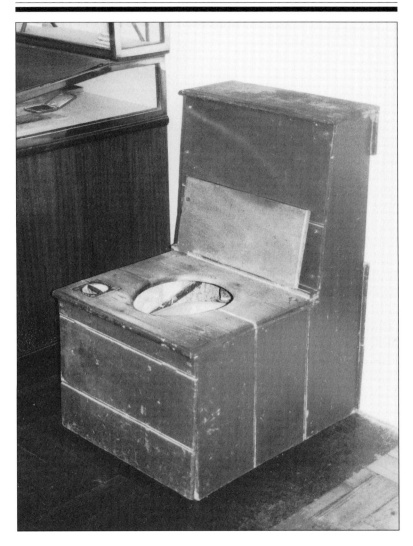

Dorset County Museum's specimen of a 'moule' has its operating handle at the front. Opening the lid on top of its back allowed fresh earth to be added to the hopper.

Moule's earth closet wasn't the first on the market but it was cheap, simple and soon recognised as the best. It also filled a desperate need for urban toilet facilities as local authorities struggled and failed to meet the demand for water-based systems.

The earth closet was a bucket beneath a hopper filled with dry earth. Ashes could be used but were not as effective at deodorising the effluent or absorbing its nutrients. After each performance, pulling a handle automatically took a measure of earth from the hopper and chucked it on top of the bucket's contents.

Provided not too much liquid was added (men in particular were expected to pee elsewhere whenever possible), the resulting mixture might be removed, dried and re-used in the hopper several times without, it was claimed, becoming offensive. It all sounds pretty disgusting but in fact a Medical Officer of Health remarked approvingly that during the process the bucket's contents were so completely changed that 'their original character cannot be recognised, and even if paper be mixed with them this disappears at the same time'.

Some closets were situated indoors, with the hopper conveniently outside the building and feeding through a hole in the wall. Some had operating handles at the back, others at the front. There were even automatic models such as those described by a naval officer in 1868: 'We use Mr Moule's closets, and employ a self-acting apparatus so that the moment a person arises from the seat a pound and a half of earth is thrown in. The closets we have established serve 300 people, and we expect to do all this with the services of a man and a boy'.

Most 'moules', as his closets became known, were bulky, fixed appliances but there was also a portable 'self-contained' model which was advertised as being 'invaluable in sick rooms, in hospitals and in infirmaries'.

The earth closet became accepted as the most satisfactory and

Another type of 'moule', with a back handle. Its wooden casing has been removed to show the metal hopper, and through its seat hole can be seen the chute from which earth fell into the bucket.

hygienic alternative where water supplies or drainage were inadequate. However, a writer noted in 1900 that it was 'very difficult indeed to get cottagers properly to use it'. Perhaps they could not be bothered with the rigmarole. Dry earth was not particularly easy to come by and it has been estimated that a Victorian family of six might use 50 hundredweight (2,540 kg) a year. People dried it in their ovens or in front of the fire, and little stoves went on sale, specifically for drying earth.

Despite the problems, Moule's closets were a success. The Earth Closet Company licensed a number of manufacturers to produce them and they were sold all around the world, to be installed in private homes, schools, barracks, prisons (776 at Wakefield Gaol, we are told), shipyards, factories and other large establishments. Queen's College, Cambridge, had moules, as did 800 people on the Rothschild estates. Moule was

Apparatus on Bearers ready to Fix.
Deal Seat 3' 0" Long.

No. A1724. " Pull Out," as drawn.
No. A1725. " Pull Up " Pattern.
No. A1726. " Self-acting " Pattern.

Strong, Portable, Self-Contained Set. Plain Deal. Galvanized Fittings. Pail complete. 21" Wide. 27" Back to Front.

No. A1727. " Pull Out."
(as drawn)

No. A1728. " Pull Up "

Strong, Portable, Self-contained.
Best Plain Deal.
Fittings of Galvanized Iron.
With Pail complete.

No. A1729. Self-Acting. 21" Wide.
27" Back to Front. 36" High.

No.			
A1724	..	..	57/6
A1725	..	..	70/-
A1726	..	..	100/-
*A1727	..	..	72/6
*A1728	..	..	86/6
*A1729	..	..	102/6

* Pails included.
Other Pails **3/7** Each Extra.

This ironmonger's catalogue of 1936 confirms that Moule's closets were in demand well into the 20th century. The self-acting model (bottom) had no handle but automatically chucked earth into its bucket when the user arose from the throne.

even awarded £500, quite a substantial sum in those days, by the Secretary of State for India in recognition of the benefits his closets brought to that country.

He fruitlessly campaigned for earth closets and recycling to be the norm right up to his death in 1880. The WC won, but British ironmongers continued to sell moules until at least 1940 and some were still in use 100 years after Henry took out that first patent in 1860.

Thomas Hardy was associated closely with the Moule family. He based Angel Clare's parents in *Tess of the D'Urbervilles* on Henry and his wife, and his actions during the 1854 cholera outbreak were Hardy's inspiration for *A Changed Man*.

An unusual encounter with a moule was described to me by Mrs Mammett of Broadmayne. 'We rented some stables in Dorchester about 65 years ago. In the loft we found a wooden lavatory with a sort of box at the back. Having nowhere for a broody hen to hatch some duck eggs, we put her in the lavatory. I saw a handle on the back, moved it and showered the hen in earth. It did not put her off and she hatched her ducklings. I am now 80 years old but have never forgotten what happened.'

[5]

PRIVY PLEASURES

The privy could be a pleasant enough place, by all accounts. Many people enjoyed sitting there on a summer's day or in the cool of springtime as they awaited events, reading perhaps, or singing quietly, watching a spider spin its web, or simply relishing the peaceful sounds of nature. They even derived pleasure from the chores associated with the privy. Here are just a few of the warm memories Dorset folk have shared with me.

Phyllis Pereira of Bradford Abbas spent her childhood in an old farmhouse in the Blackmore Vale, where the privy lay 'down a garden path that was stone flagged all the way, with a flower border on one side and a low box hedge on the other. Snowdrops grew in abundance beneath the hedge. The building was about eight feet square, brick built with a pitched roof over which ivy grew in profusion, rather like an umbrella. The floor was of wood and the walls were whitewashed inside, with a tiny window in one wall. The *pièce de résistance* was a two-holer seat.'

At the age of three, Mrs Sandra Banks was evacuated from London to a thatched cottage at Witchampton, near Wimborne. She has happy memories of oil lamps, water from the well and, in the garden behind the house, a privy with a well scrubbed wooden seat. 'That little hut was a haven where I could shut myself away from my brother and read my Enid Blyton books. I also read the small squares of newspaper strung up on a hook on the door. If an adult was in the hut, you could see their feet under the gap below the door. It was a favourite place for them to read the paper.

'My brother was too scared to go out after dark. I have memories of a torch and insisting on an adult to accompany me against the perils of the night. The large bucket was emptied

A delightful little stone privy in the colourful garden of a 17th-century cottage at Beaminster.

into a large pit at the top of the garden, then covered in white powder. We had beautiful vegetables from that garden all year round.

'At the village school, the girls' lavs with three holes in a row became a gossip shop at playtime. The strong smell of Jeyes' Fluid still brings memories flooding back.'

As a lad, Mr Rawles of Wareham was usually to be found at his Gran's house. 'We called her outside loo the Thunderbox. It was my job to tear the newspaper into squares, make a hole in one corner for a loop of string and then hang it on a nail on the inside of the door. When we were doing our business, we used to read a bit of news on the top square of paper and then try to find the piece with the end of the story. We rarely succeeded.'

'I grew up in Parkstone and we were blessed with an outside loo,' says Joan Davies. 'I well remember having to cut up

43

squares of newspaper, making a hole in them for the string to go through, then going out and hanging them on a nail. It was great fun sitting out there (when it was warm weather!) trying to piece the squares together to make a story.

'The loo outside also became a place of refuge for me, and if I had done something that I should not have done and I knew that I would be in trouble, I would lock myself in and hope if I stayed there long enough my mother might forget all about me. But it never worked, for sure enough after a while I would hear Mother banging on the door of the loo, shouting, "It's no good my girl, you have to come out sometime!" And so I did, and had to face what was coming to me. Oh, what days they were.'

As a girl growing up in Allweston, near Sherborne, Jean Gaster disliked walking and rode her bicycle everywhere, giving her Gran the opportunity to remark, 'That maid would ride her bike to the privy if she could!'

'Our privy was a short path away from the house, passing sweet smelling honeysuckle and pink everlasting sweet peas,' says Jean. 'I remember making squares of *Radio Times* threaded with string – one up from the daily newspaper or *Western Gazette*!'

Maureen Linnell recalls a much nicer alternative to those newspaper squares. 'My mum worked in a greengrocer's and whenever a box of oranges had been emptied she would bring home the tissue wrappers. I spent many winter evenings smoothing out orange wrappers for use in the loo. Not only were they softer than newspaper, but the odour of oranges was pleasant.'

Mrs M. Downes now lives in Poole but she was raised in Wimborne. 'From the age of ten it was my Saturday morning job to clean the privy,' she says. 'My Dad used to empty the bucket in a pit in the garden on Friday night after dark and disinfect it with carbolic powder. I had to scrub the seat with hard carbolic soap and whiten the concrete floor with a hearth stone, then a new sack bag was put on the floor.

Miss Frances Woodward kindly showed me these former wash-houses, which stand in the communal yard of a terrace of estate workers' cottages (1864) in Puddletown High Street. Built into the back of each wash-house was the cottage's privy.

'All the waste went back into the soil and we had lovely organic vegetables – no problems with insecticides in those days.

'We used to fold the newspaper into squares and run a knife down the folds. When we had a nice wad, we would push a piece of wire through one corner and tie it to a nail which had been hammered into the privy wall.

'Some people used to hang pictures in their privies. For some reason, my grandmother always had a calendar on her wall.'

Joan Chambers from Winton had never experienced the delights of the privy until she began attending school. 'They were bucket toilets in a creosoted building outside the school, with pieces of newspaper hanging on a nail. I remember them well. I hadn't seen anything like it before. Then, when I was first married, we lived with my father-in-law in his cottage at

Corfe Mullen. He had a privy at the very bottom of the garden, an earth closet, and by the privy door was a trench where he grew red and black currants, besides lots of rhubarb and plenty of other vegetables.'

Just before her 91st birthday, Mrs Elsie Webber of Weymouth shared with me these extremely lucid memories of her childhood. 'I was born in Langton Matravers in 1907, the youngest of ten children. Although I was the weakling of the family, I've outlived the rest of them.

'We lived in what must have been a farmhouse. It had four bedrooms, a sitting room, a living room and a large back kitchen, where we had a copper and a mangle and a big, long, galvanised bath, which hung on the wall with a very big hook.

Twin privies, back to back, were the facilities for two cottages at Bagber. A former tenant, revisiting in her 96th year, told the present owner how she used to sit side by side with her aunt on the two-hole seat.

We burnt logs to heat the water in the copper. It was on all day Friday and Saturday for us to have a bath and wash our hair, then again on Monday to do the washing. My poor Mum was all day doing the washing but I never once heard her grumble.

'We had a big tank in our garden and we used this rainwater for washing and for baths but not for drinking. Water for drinking came from stand pipes. We were lucky to have one just across the road but some people had a long way to walk with big buckets of drinking water.

'Our privy was built onto the back of the kitchen, with the farm stables on the other side. My Mum always had flowers along the front of the privy, always nasturtium – she loved them. The privy had a smooth wooden seat with a big bucket underneath. There was a potty on a little stool in the corner for the children.

'At the bottom of the garden was what we called a mixen, where the privy bucket was emptied and then covered with potato peelings, cut grass, garden rubbish and so on. When everything was rotten and getting a bit smelly, Billy Wingate, the farmer, came to empty it. He had a big greenhouse and grew tomatoes. They were lovely, I can still remember the taste. Everybody loved them and they came from miles around for Billy's tomatoes.'

Hilary Collier was brought up in Sandford Orcas. Their privy was surrounded by montbretia, which were always referred to as lavatory flowers, as they are to this day by certain members of her family. A pleasant surprise awaited her when she set out to trace the cottages at Silverlake, near Sherborne, where her great-grandparents had lived after marrying in 1860. 'Much to my amazement, in an area overgrown with stinging nettles I found not only the ruins of the cottages but two double privies, still standing and with a corrugated iron roof. Each one had a wooden seat on two levels, the higher level with an adult sized

In the grounds of Cannings Court at Pulham stands this well-preserved two-hole privy, a very long trek from the farmhouse it once served. The drystone wall at the rear surrounds a brick-lined vault, sufficiently large and deep to hold the accumulations of many months, if not years, between emptyings.

hole and the lower with a smaller hole for children. The family now refer to it as mum's ancestral double loo!'

Mrs Collier also remembers an elderly spinster, Lottie Lane, who lived in Ryme Intrinseca in the 1940s. 'Her privy was at the bottom of the garden and its walls were covered with pictures from old magazines and calendars. Sitting there, one faced three large pictures, copies of oil paintings with their frames removed. The first showed a young man lighting a clay pipe, in the second he was apparently enjoying the smoke, but the third made one realise that this must be his first experience of smoking. Pipe no longer in use, he was clearly feeling very sick.'

Mrs Margaret Bamford recalls the holidays she and her girl-friend spent at South Road in Bridport. 'It was the end cottage of a row of six. Down the garden were six privies in a row, each one

The two-hole seat inside the Cannings Court privy could well be the original. Note the removable lid – all the best privies had some form of cover to reduce the nuisance from pongs and flies. A former land girl remembers often sitting in here while working on the farm during the war.

in the position relating to a cottage. As we were in the end cottage we had to use the end privy. There were no lights, and I remember my friend always asked her boyfriend to take her down to the loo when it was dark as she was too nervous to go on her own. I suppose she thought I wouldn't be much protection. Those were happy days, though.'

This lady asked me not to use her name. She lives in Wool and had used an outside privy for 70 of her 88 years. She writes: 'I can well remember having to cross the back yard, whatever the weather, to pay a visit to the loo, which was built into a corner of the wood shed with a separate entrance for privacy. It was built of rough timbers sawn from the outer part of a tree. Inside was a wooden top with two holes, one for an adult and a smaller

A block of five little outhouses at the bottom of their gardens served five thatched cottages in Sturminster Newton. The doors discreetly face away from the cottages. Situated in a conservation area, their future seems secure.

one by the side for a child.

'A box of lime, sand or ashes was put nearby, with an old fire shovel to sprinkle into the bucket to cover what today is washed away by water.

'In those days, the interior of a loo was either whitewashed (a mixture of lime and water to the consistency of paint) or papered with any oddments of old almanacs or newspaper, as wallpaper was either not available or very expensive. I lost my mother at a very young age and was looked after by my grandmother, as was my father and brother. She was quite enthusiastic and used to paper instead of whitewash.

'During the winter evenings my brother and I were occupied in cutting squares of newspaper, tying so many to a bundle on a piece of string to hang in the loo for use. What with the squares of

Glorious views across Christchurch Harbour to Hengistbury Head were the reward for visiting this privy behind a former fisherman's cottage in Mudeford. Not much fun when a winter westerly was blowing, though. You can guess where the bucket was emptied.

newspaper, and newspaper papering the walls, one could sit there and read past news.

'I was terrified of spiders, which one could hear crawling behind the paper, so I used to get my brother to come with me to stand outside the door. I would sit on the loo and bang my feet against the woodwork to frighten the spiders away. Then out would come Granny, telling me to be quiet as the neighbours would know where I was.

'They were happy days in spite of the hard work. Unhygienic perhaps, and maybe appearing to be unhealthy, but no harm came to us. I have now been living with the use of a flush toilet and hot and cold water. Maybe life is easier and convenient but I'm inclined to think it's no better healthwise.'

[6]

PRIVY PERILS

Although many people derived great pleasure from their privy and look back on it with genuine affection, there was a darker side. Owning an outhouse brought with it a whole raft of risks. It paid to keep your wits about you while performing your daily duty, as all manner of things could – and did – go wrong.

Irene Webb of Winterborne Stickland remembers that rats often came out to sit alongside her in the privy of their former home at Winterborne Clenston. Her most vivid memory, though, is of a winter's day when, perched innocently on the seat at the age of just five or six, she suddenly slipped down through the hole, straight into the bucket and its unwholesome contents. 'My grandfather, bless the dear old gent, washed my bum with rain water outside the back door,' she recalls.

Other things besides little girls sometimes disappeared down that uninviting hole. Marjorie Dawkins tells of the day her aunt visited a pub and had to use its two-hole privy. 'She went in and sat down, put her handbag on the seat beside her and it fell through the spare hole into a full bucket. The landlady had to call her husband to get it out. He was not pleased, and after fishing the handbag out with a stick he handed it to her, just as it was.'

From R. G. Smith of Bournemouth comes the tale of his family loo, which sat in an orchard 50 yards from their cottage. 'In the orchard we kept some chickens with a cockerel. In the loo was a bucket which was emptied into a trench in the garden, and the job was done by grandad. There was a trap door at the back of the loo to take out the bucket for emptying. On this particular occasion, grandad couldn't have shut the trap door afterwards. I was about seven years old, sat on the seat and after just a few

Ventilation louvres in its side wall assisted in keeping a relatively sweet atmosphere within this bucket privy at the Manor House in Bagber. The drawback was that it is set in a little wood, quite a trot from the house for anyone in desperate need.

seconds something bit me in the rear. I smartly jumped off, faster than I had sat on, to discover that the Light Sussex cockerel had somehow found its way into the empty bucket. I ran up the garden path as fast as I could, in great distress, with the affected part bleeding. Everyone laughed when I told what had happened but I'm pleased to say that first aid was rendered.'

I met a woman in Long Crichel who cannot forget a traumatic incident involving the stout timber privy in their cottage garden. Each morning she threw the ashes from yesterday's fire onto the garden, until one day her young son came rushing indoors to say that the hot cinders had set their privy on fire. Flames were licking up the back wall by the time she got to it. The only source of water was a well in the lane outside, from which they frantically hauled up buckets of water to throw onto the blazing privy. It was a long job and her abiding memory is of seeing the ample contents of their privy pail actually on the boil by the time the flames were extinguished.

Hazards associated with the burying business were inclined to be particularly unpleasant. Mrs Challenger reminds us of a regrettably common incident, on this occasion concerning her grandparents' privy which was at the back of their farm cottage, surrounded with bushes. 'My grandfather was responsible for emptying the bucket but there came a time when he could no longer do this without assistance. One day my father was due to go and help him with the task but as we lived some miles away grandfather got fed up with waiting. He decided to do the job by himself but whilst carrying the bucket he slipped and fell, bringing the contents all over himself so that, as father informed us on his return home, "Grandad was covered in shit."

'At my grandparents' cottage the water had to be drawn from the well and then boiled in large kettles on the big open fire, so getting him clean would not have been an easy task.'

An equally traumatic tale comes courtesy of a Mr Kemp, who wrote to tell me, 'My first experience with privy middens was an unfortunate one. This occurred when, as schoolboys, my brother and I were invited to spend the day with two school chums at their parents' home. Their father was a carnation grower, owning two nurseries. The one we were shown around on this particularly unforgettable occasion employed a large staff and although of course flush toilets had been in use for a considerable time, this nursery had not changed over from privies to accommodate the workers, so there was an ominous and malodorous row of them in evidence.

'The effluent was channelled off into a large, open sump pit, which had a covering of straw and branches to conceal its presence – too well, in fact, since on traversing the adjoining footpath yours truly managed to blunder into it, feet first and up to my thighs. Fortunately I managed to blunder out of it quicker than I fell in, amidst some thankfully controlled mirth from the bystanders. The finish to that day was me being washed down with a hosepipe.'

The privy should have been a place of calm and sanctuary, somewhere to forget your problems, to daydream or simply to concentrate on the job in hand. Young George Simpson, though, rarely enjoyed such pleasures as he clung desperately to the seat. 'We lived in the country, not far from Cerne Abbas, with a privy built over a stream that ran down the boundary of our father's land. It was actually built out into the stream on wooden legs. There was just a hole in a seat inside and it went straight into the stream. Other people had a similar arrangement so the water in that stream must have been really healthy! Mosquitoes used to buzz around in the summer and bite your bum if you stayed too long. Horse flies were worse, you knew all about it if one of them took a bite.

'My two older brothers thought it great fun to torment me.

Each of the old coastguard cottages at Mudeford had its own privy, back to back with next door's. Bashful users had to contend with the fact that they were just a few yards away from the front windows of the cottages, in full view of the neighbours.

They would wait until I was sitting comfortably before hurling stones or bricks, the bigger the better, into the stream just behind the privy. The aim, which unfortunately they succeeded in achieving more often than not, was to splash up as much water as possible onto my exposed and defenceless backside. Sometimes they forced me to retreat but on many occasions the needs of nature demanded that I stay the course, bravely performing my duty to the accompaniment of a barrage of bricks and vulgar barracking. I like to think it made a man of me.'

Another person who will never forget her ordeal in the privy is Audrey Hall. Her story begins with her husband's demob from the Army, when he solved their accommodation problem by taking farm work with a tied cottage in the little hamlet of Pentridge. 'There were two side-by-side cottages up the hill by the

At the time I photographed this unusual arched privy in the ancient hamlet of Pymore, just outside Bridport, it was due for renovation as part of a regeneration programme and is in fact a listed building. I met a former resident who remembers it in use, just a small room with seats high above the stream. Everything splashed straight into the water, in full view of other villagers.

church,' recalls Audrey. 'Number One was ours, with a stone-flagged kitchen, cold water tap and a kitchen range for warmth. Outside was just a small square building with a corrugated iron roof, a square seat with a hole and a large bucket. Having never had to cope with outside plumbing (or in this case even less!), it presented many a dilemma but we took it all in our stride – in fact, to the large pit at the end of the garden which in time was filled in and a new one dug as required.

'However, nothing had prepared me for the visit of Lord Cranborne's shooting party one day. His estate keeper sent the beaters up our hill to flush out the birds from the fields behind our home. They flew straight over our privy, to be immediately peppered with lead shot from the guns. Unfortunately, I hap-

This picture shows how some privies were dark and claustrophobic. I have heard many tales of well-endowed people finding it impossible to turn around once inside. Instead, they had to reverse into the privy and onto the throne.

pened to be ensconced in the privy at the time, all unawares, and the noise of the shot rattling on the corrugated iron not only put the fear of God into me but also drowned my calls of surrender from within. I was forced to stay in that unsavoury place for some time, until thankfully men and guns moved on, never realising the other "bird" they had almost bagged.

'I did have thoughts of hanging a white flag through the moon-shaped hole in the privy door but with my nether garments being the only thing available I thought discretion the better part of valour.'

I have included this picture from Lytchett Minster to show how spooky some privies could be. Imagine coming up here at night! Not long ago, a passer-by reported hearing heavy breathing from within. The owners cautiously investigated, to discover an owl had taken up residence. They closed the door and left it in peace.

59

Finally in this chapter of dunny disasters we meet Elizabeth Rose of Christchurch. She has good reason to remember the pretty privy of her childhood. 'It was at the bottom of the garden, smothered in old man's beard and roses, and screened on one side by high bracken and in front by trellis. Imagine if you can a small child sitting there, swinging her legs which didn't touch the floor. On looking down at my feet, I saw an adder emerging from the ventilation hole in the metal base of the throne. I screamed and screamed and ran, to be doomed to a lifetime of constipation.'

[7]

PARTICULAR PRIVIES

It wasn't strictly necessary to site your privy way down the path. Some people arranged matters so they could enjoy paying a visit without going outside in all weathers. Earth closets were often installed inside the house, any nasty niffs being smothered by the coverings of dry earth or ashes. At least, that was the theory.

Maureen Linnell knew an indoor privy at Fordington, Dorchester, that was situated in a narrow passageway between back door and kitchen. 'It was regularly whitewashed, had a polished tiled floor, was lit by an oil lamp because there was no window and had a heavy wooden double seat. By use of much disinfectant the odour was reduced to a minimum. Still, the eventual conversion to a WC was greeted with more than relief.' I bet it was.

The only truly safe way was to expel the stuff from the house as quickly as possible. Mr and Mrs Bruce Pye are justifiably proud of the novel indoor privy at their 19th-century home in Beaminster. Accessed from within the cottage, the tiny room overhangs the River Brit, so when it was in use everything fell straight through the hole into the river and was swept neatly away.

Most people far preferred to keep their privy and all its works well away from the house. Sometimes a great deal of ingenuity went into the design of a particular privy, resulting in some interesting varieties.

At Bill Chitty's boyhood home, for example, there were two steps up to the boxed-in loo, which must have been like mounting a throne. 'On the side was a container of sand with an arm chute which you turned round and put sand into the bucket to cover the contents. When the bucket was full, it was put in a trench in the kitchen garden.

61

Who could fail to be impressed by this splendid specimen! Part timbered, part herring-bone red brick, it adjoins a 400-year-old cottage at Tarrant Monkton. The owner says it was a two-holer, and there is an emptying hatch at the end furthest from the cottage. Truly magnificent.

'A lot of big houses provided their gardeners with a wooden shed with a built-in seat and two handles front and rear. It was placed over a large hole in the ground and when that was full the shed was moved to another. Others just had a piece of guttering which went through the privy wall onto the compost heap.'

Bill also recalls an incident when he and his friends tried to take revenge on another apprentice. 'We waited until he went to the privies, then set light to an oily rag, lifted the bucket flap and put it in. But it was the foreman and out he went in a hurry – he thought his pipe had set light to his coat.'

R. J. White remembers the privies on the Forde Abbey estate, where Dorset meets Somerset. 'My mother and I went to stay with her friend who lived in a cottage on the estate. The privy was a ghastly place in a hut some 20 to 40 yards from her cottage. However, as there was no sleeping accommodation for me I had to sleep at another cottage. The loo there was almost modern. It was situated over a stream which was the overflow from a lake in the grounds. As there was a drop of four or five feet from the seat into the fast flowing water, it was quite hygienic – except for the fact that it went straight into the River Axe, untreated. A particular problem arose when the Axe was in flood. Then it was a wet journey to relieve oneself.'

Aubrey Jenkins lived in old Poole as a boy, where a pair of houses shared one outside tap and a privy (they called it a 'dunnegan') at the bottom of the garden. 'Actually, it was partitioned so there were two enclosed wooden seats. If there were

This remarkable grade II listed building is the privy of a Regency farmhouse (circa 1815) at Long Crichel. The entrance is a fine arched doorway on the other side. It could be that originally a pit beneath the seat emptied through a small arched aperture (near the wheelbarrow) directly into the farmyard. The hatch is probably a later addition, after conversion to a bucket privy.

This shows the other side of the emptying hatch of the Long Crichel privy tower. Its bucket is missing. The thoughtful seat design offers three different holes to accommodate various sizes of backside. Clearly, this whole set-up was designed for someone who demanded a certain refinement.

two occupants and a bit of flatulence, most interesting sounds reverberated through the building. For night-time use we had the usual ceramic bowls under the bed. All urine was emptied on the garden, as my grandfather swore by it for the growing of parsnips, and he grew some really good ones.'

An ingenious arrangement from his boyhood is remembered by George Samways of Portesham, near Weymouth. The village school adjoined a river, into which a hatch had been set. When necessary, some of the river water could be diverted into the school grounds to flow through a 12-inch pipe. It ran beneath first the boys' and then the girls' straight-drop privies. This stream of piped water then continued down the full length of the school, collecting along the way any offerings splashing down from the family toilet in the adjacent schoolhouse, before depositing its fragrant load into the same river further down-stream. 'A good system in those days!' says George. No doubt it lent a soothing quality to the experience to hear that stream of water gurgling gently beneath as one meditated.

Another example of how people strove to improve the basic privy was discovered by Mrs Buncombe of Bournemouth when she and another Army wife rented a house with a loo at the end of the garden. 'It had a box-like seat, quite high, so it was a long drop to the water flush, which was a sort of metal trap door. When the water reached a certain level the trap would turn over, making a terrific noise. It didn't take my friend and I long to work out that if we left the kitchen tap running, we could make the trap suddenly turn over noisily when someone was using it. I'm afraid we usually waited until our lady friends were visiting!'

Youngsters who only encountered a privy when staying with older relatives were often fascinated by the novelty. Janet Clarke loved visiting her grandmother's farmhouse near Porte-sham in the early 1960s. Her back garden privy was a stone building with a corrugated iron roof. Janet recalls: 'There was a welcome three-inch gap at top and bottom of the rough wooden door, which helped ventilate and illumine the dark recesses of the tiny interior. There was just enough room for a wooden bench seat across the back wall with a round opening onto the concealed bucket. Beside this was an old cup and a large tub of lime powder which one used to scatter on top of the latest deposits after the mission was accomplished. On the inside of the door Gran had hung a notice: "Don't sit there all day dreaming of that £75,000!"

'The bucket was emptied frequently into a cesspit buried at the far end of the very long garden, which in turn was emptied by a specialist who called with his container vehicle every few weeks or so.

My picture cannot do justice to John Enderby's fascinating wooden privy at Fontmell Magna. The floor is an engraved headstone (Edward Spicer 1857), presumably purloined from the churchyard opposite, and the rail over the door is said to be from the burial yard of the Methodist Chapel. It bears a partly decipherable inscription 'Samuel son of Isaac . . .'

'My grandmother ran a cream tea business and my cousin and I, who helped to serve, were always amused when visitors asked to use the "ladies". We would tell them to follow the path round the back of the cottage and they would come to it. Their faces were a study on their return.'

Mrs Frances Cude of Swanwick, near Southampton, enjoyed many August holidays at her grandmother's house in Shipton Gorge. They were marred only by the sanitary arrangements. 'Gran's back garden was up a bank and the loo was at the top. The seat was scrubbed almost daily and newspaper squares hung on a nail. Of course, we had a po under the bed but frequently the call of nature was a greater need, so at night Gran would have to light a candle to escort us up the garden path. I hated the massive spiders that hung in the corners. The whole ordeal was a complete trauma but we needed a holiday and to suffer for a month was worth it just to see Gran. I have no idea how they coped with emptying the thing but the memory is still fresh in my mind.'

Mrs Hutchins from Broadmayne grew up with a privy down a long garden path. 'It had a pit where my father used to put more soil in every so often. My brothers had to take me there when it was dark, and they did not approve. They had to take it in turns to go with me with a candle which sometimes decided to go out – we had no torches in those days. They would wait for me outside, grumbling, "Have you done?" or "Hurry up!" and often gave me a clout when I eventually came out after taking what they thought was too long.

'We used newspaper cut into squares, which always had to be done Saturdays in case we had visitors on Sunday.

'After I married, we still had an outside loo and my husband had to dig a hole in the garden to empty the dreaded contents of the bucket. He being a herdsman, we moved around and in 1952 we came to Sturminster Marshall which to my horror still had

In 1992, about £1,500 was spent on expert restoration of these privies behind farm cottages at Chetnole. The final touch was providing catalogues from Harrods and Fortnum and Mason instead of newspaper squares (unfortunately, they were soon destroyed by mice). Then there was a grand ceremony with champagne. These must be the only privies to contain a plaque certifying that they were formally 'opened' by the local MP.

an outside loo, although it was set in the shed so it was a bit more private and comfortable. But I still had to scrub the seat clean with the washing water from the Monday washday. We didn't have to empty that one, as a big tanker used to come every Friday to do it for the price of one shilling. It was well worth it.'

W. E. Wilkinson attended Sunday school in the 1930s at Throop Chapel, where the men's conveniences consisted of a communal bucket beneath a wooden seat. Their urinal was simply a lead trough filled with peat. At least the vicar's family benefited from these arrangements. 'The caretaker always emptied everything into the vicarage vegetable garden,' remembers Mr Wilkinson, adding helpfully: 'The caretaker was Mr Chard but we called him Flapper Chard because he was small and waddled along like a seal.'

Another stout pair which have seen better days. These are at the back of farm cottages in Upwey. The left-hand door is probably the original, judging by the top ventilation gap.

[8]

THE HAPPIEST DAYS OF YOUR LIFE

At the age of five, Pat Lowe – or Pat Conway as she was then – began a seven-year acquaintanceship with the wooden-seated privies of the village school at Wimborne St Giles. 'Graffiti was rife in those days, as in all loos,' she says, 'but it was all very innocent, unlike what is seen in ladies' toilets these days.' By the time she left, in 1953, Pat had been tempted to add her contribution to the privies' whitewashed walls, 'albeit only drawing a pierced heart and writing "Pat Conway loves Michael Scriven". My friends all wrote similar things. It was in pencil, there were no Biros in those days, and we were made to spend our playtime cleaning it off the wall with rubber erasers.'

Her aunt was born in the village in 1912, when it was known simply as St Giles, and she attended the school 25 years before Pat was born. 'To think my niece sat in the same privy, on the same seats as I had so many years before!' Kitty Sinnott told me. 'I remember they were along one wall, three or four of them in little compartments, each with its own door. The boys' privy was in their playground. The privies were kept well scrubbed but there were deep pits below, not emptied all that frequently, and my mother used to say you could smell from a child's clothes if she had been sitting on the seat too long.

'We lived next door to the school and it was my father and one of his workmen who went at night with wheelbarrows and shovels to empty the privy pits. They wheeled it to a garden next door to the school and there the wheelbarrows were emptied. This garden belonged to my uncle and it was kept in beautiful order. What lovely vegetables! They certainly tasted better than vegetables do today.

'My mother was an infant teacher at the school for over 40

70

Most schoolchildren had to cross the playground in all weathers to use out-door privies. These ones are at Lytchett Minster.

years. From the age of four, myself, my brother and my sister were all in her class. She always kept a small pair of trousers in her cupboard in case a little boy should have a mishap. I don't know how the girls fared.

'I can well remember that if some child made a dirty smell in the classroom, my mother would demand, "Who has made a dirty smell?" Usually no one answered. Then she would roll up some newspaper, light it, blow out the flame and go around the room, swishing the smoke around.'

Not having a privy at home, Linda Sanders experienced something of a culture shock on school days. 'I attended Salway Ash Primary School, near Bridport, from 1957 to 1959, where we had outside loos which were nothing more than a shed extension to the school. There were large metal buckets with wooden seats

71

on top. We had no running water; a pump in the playground
provided water for drinking at lunchtime and for washing our
hands beforehand. We had to stand in line, one of us children
having been given the Pump Monitor job of drawing some
water from the pump and putting it into a bowl for us all to
wash our hands. You can imagine the state of the water if you
happened to be low down the queue. Hardly hygienic, although
obviously the attempt was there to instil the habit of hand washing.

'We hated the loos – of course, we called them toilets. They
stank! Once a week they were emptied by council employees
who we called "the stink men", poor things. They wore brown
overalls and had the unsavoury job of carting the buckets to their
lorry, which contained a tank, emptying them and returning
them with a "healthy" amount of something like Jeyes' Fluid,
which at least meant a fairly nice smell for a day or two, until it
was overcome by the new contents.

'When I say we hated the toilets, I think perhaps I was more
sensitive than most. Some of the children came from rather basic
homes, we were nearly all from farming families and many lived
in somewhat squalid cottages. I did not.

'Like all children, we tended to hang around the toilets for the
usual naughty chit-chat, illicit smoking and so on – all anathema
to me, though I had to join in, not wanting to be thought a
snobby outsider. My problem wasn't the naughty activities but
the smell.

'The school itself was absolutely wonderful, with marvellous
teachers who handled the very basic conditions admirably. I
consider myself lucky to have been there, but I will never forget
the stink men either. I suppose they were not unhappy to lose
their odious task when the school was modernised, sometime in
the 1960s.'

Buckets and wooden seats were also the order of the day at
Melbury Osmond School, as in most houses in the village, when

Patricia Lee was there in the 1940s. Just two buckets in the yard behind the school served the girls' needs but at least Izal squares were provided instead of newspaper. Matters didn't improve much when Miss Lee moved up to Evershot School in the 1950s – three buckets for the girls, emptied at weekends, while the teachers enjoyed the luxury of a flushing WC just a few feet away.

A number of former Hampshire villages have now been placed in east Dorset, which prompted Mrs Challenger from Christchurch to tell me about the little school at Burton. Nowadays it's for infants only but everyone went there in her day, and the privy at the bottom of the playground came under particularly severe pressure during the war years.

'It was a small brick building, one toilet for the girls, one for the boys and another for the teachers which backed on to the others and faced the school. There were just three teachers and three classes but I would guess that each class had 30 children during my time, as we had evacuees from Bitterne, Southampton. The only wash basin was a small one in the lobby attached to the large classroom and I am sure most of us would have forgotten any hygiene by the time we had raced back from the end of the playground.'

It also sounds as if crossed legs were commonplace. 'We didn't ask to be excused during lessons unless in dire need,' says Mrs Challenger. 'You would have felt very conspicuous, and in bad weather you would get soaking wet just getting there and back. Almost all of us went home for lunch, a very long walk in many cases, of which I was one, so we didn't have time to visit the school privy. The most important thing was getting home. Then at the end of the day you made a bee line for home, you never hung around school.

'I believe they were connected to the main sewer when it came through the village in the early 1960s. The school and playground look just the same to this day but the privy building is no longer there.'

Built like a brick outhouse! This 'fortress' was the privy block of the tiny school at Winterborne Monkton. Boys entered through the arch and stood in the rain, as the twin roofs slope inwards without actually meeting. The girls' entrance is at the other end. Buckets were removed through three hatches at the rear.

Other privies that apparently were best avoided were those at St John's Church of England School in Wimborne Minster. Cherry Smeeth tells me that her son and daughter, whose school it was, would 'race home at breakneck speed to avoid using its bucket loos'.

Even the more refined systems seemed designed to produce an aversion to using the toilet. From the age of three, Doris Boswell-Smith attended elementary school in Broadwindsor, in the far west of Dorset. Pupils were spared the full horrors of the primitive privy as there was plumbing of a sort installed. What was lacking was privacy. In a small room with no separate cubicles, five or six girls sat side by side (cheek by cheek?) upon round receptacles that jutted out like so many upturned pipe bowls. An enormous drain disposed of matters. This wasn't the place to be bashful or overly modest, and perhaps joining your little friends made the business less of an ordeal.

A certain degree of courage also seems to have been demanded of pupils attending Swanage Infant School, where, according to Maureen Linnell, there was 'a system handed down from the Romans.'

This is what she recalls: 'A row of loos had been built against the playground wall. Although divided into cubicles, the seat was just a continuous length of planking. Beneath it ran a stone channel. A cistern at one end emptied itself at regular intervals, the water swooshing down the channel with such ferocity that new pupils – and even the more nervous older ones – would shoot out of the loos like corks out of bottles. The system had worked for more than 50 years when I left Swanage in 1955.'

So we discover ample evidence that children the length and breadth of Dorset would bottle it up for hours rather than face the rigours of the school privy. Little wonder, then, that constipation was an ever present concern of anxious parents. It's one reason why children in those happy years faced yet another dreadful ritual, the obligatory purgative dose to keep them 'regular'. There are grown men who still shudder at the memory of joining the family line-up to receive a mighty spoonful of castor oil, whether they needed it or not. Senna pods and syrup of figs were other favourite remedies, while many mothers contrived to serve rhubarb at least once a week. You couldn't be too careful where bowels were concerned.

Whatever the drawbacks of the old school privies, at least they were the source of much innocent amusement for small boys. Many are the tales of the pranks they played upon little girls who, believing themselves safe in a place of ease and privacy, instead were subjected to all manner of indignities.

75

From Mrs Hutchins of Broadmayne come her memories of the school toilets in a tin shed, four each for girls and boys. 'The boys used to go round the back and bang on the tin walls to frighten us girls, and the bigger boys would go in there for a quiet smoke,' she says.

Bucket privies with emptying flaps at the rear could have been designed for the delight of mischievous boys. They made an irresistible target. It was almost too easy to sidle round the back of the privy block, lift a wooden flap and torment the unsuspecting user within.

Even the teachers were at peril in the privy, as Denzil Perry describes in the following tale from his village magazine. 'In 1930, aged five or six, I attended St Andrew's Church School in Fontmell Magna. One of the most indelible impressions on my mind was the toilets. On the west end of the building was a cold, very forbidding greenstone toilet block – girls to the left, boys to the right. Scrubbed seats, no lids, no water, that shiny paper (if you were lucky) and the smell of Jeyes' Fluid. Down below there was a long trench-like vault, which was cleaned out by poor old Paul Clark, the church sexton, for a pittance.

'One day, we were called to assembly and given a stern talking-to by the rector, Canon Edmonds. We weren't told any facts concerning the apparent hubbub in the school but received a lecture on "telling the truth and admitting the awful deed".

'He said, "Own up now and all will be forgiven." Not a sound, not a murmur – but what was he on about?

'Months later the facts filtered down to us infants. Poor, timid little Miss Wilson, who taught the Middle class, had visited the toilet block and had received a most unwelcome bunch of stinging nettles to her posterior. The boy, who shall be nameless, must have crawled along the vault to carry out this terrible deed. He was never named but in the end everyone knew who he was.

'Poor Miss Wilson was away from school for a fortnight – but what *did* she tell the doctor?'

[9]

You'll Never Make It In Time!

Anyone who has read this far must surely admire the fortitude and resilience of those for whom a visit to the loo entailed braving the elements and much more besides. It had its plus points – a former district nurse confided that the walk to the privy was all that kept many of her elderly patients alive – but some unfortunates almost had to undertake route marches to reach their destination. For them, hanging on until the last moment could have disastrous consequences.

Mrs Rosemarie Suter of Gillingham grew up in an old farmhouse in the tiny Dorset hamlet of Queen Oak, between Bourton and Zeals. It was quite a hike to their privy. 'We had to go out our front door approximately eight yards, across the front path, turn right and then go up the hill about 25 to 30 yards to the toilet, which was actually in Wiltshire. I don't expect many people had to go to a different county to spend a penny and then back to the other to go to bed!

'It was joined on to the cow stall and using it during the war was not very pleasant as we had airmen – Yanks or Fleet Air Arm – in huts just behind the toilet. We couldn't take a bright torch with us, it had to have a shade because of the blackout. The evacuees were frightened at first but soon thought it was great fun.

'There were two wooden seats with big buckets. One seat was high with a big hole, the other was lower and had a smaller hole for children. I now feel so sorry for the people who had to empty the buckets. And, yes, we used to cut up the newspaper and thread the squares on string to hang on the wall.'

John Loader from Wimborne told me, 'My family lived down a lane off East Borough, with our grandparents living next door.

This survivor from the pre-flush days is in the small back garden of a farm cottage at Burton Bradstock.

The cottages had only front doors, with a cold water tap (the only supply) outside each one. There was no back garden or yard, so if we wanted to use the loo we had to leave the house by the front door and turn left, past our tap and the water butt which stored rain water for baths and laundry. The loo was on the left, through a door in the side of our barn, which was about two feet above ground level.

'The wooden loo seat was regularly scrubbed and gleaming white. Males had to lift the seat, which was about three feet long. It would stay up on its own provided you pushed it right back; if you didn't, it could be dangerous as the seat was heavy. Another hazard at night was boys forgetting to put the seat down afterwards. I knew one old lady who failed to take a torch and was probably marked for life in a spot she could not display.

'Our usual toilet paper was newspaper, cut into squares,

The *Daily Mail* was very popular . . .

pierced with string and hung on a nail. You could, of course, take a paper to read and use one of its pages. *Old Moore's Almanack* was favoured by some people for its many pages of good quality soft paper.

'My grandparents next door had to turn right out of their front door and into the corner of the yard alongside the house for their loo. It was a free-standing structure, purpose-built of galvanised corrugated iron, and for privacy it faced a six foot corrugated iron fence. The colour of both toilet and fence depended on whatever old leftover paints my grandfather had mixed together. He was a thrifty man and ignored grand-mother's comments about not liking the colour.

'Grandad was my idol and as a small boy my search for him often led me to that toilet. He would be sat there smoking his pipe, reading the paper with the door ajar. The seat was long enough for me to sit alongside him and chat, although he usually found a reason for me to leave before carrying out the final operation.

'I am the proud possessor still of a bucket now used for garden weeds and so on but which formerly served in other ways. They made them of heavy gauge metal, well galvanised with a stout carrying handle and others at each end of the oval for tipping. They don't make them like that any more. It will see me out.'

Christine Whipp wrote to tell me that 'the only funny family story I can remember about privies concerns my Great-grandma Tompkins who was bitten on the backside by a hornet while using her outdoor privy at Maiden Newton.' However, Christine does retain fond memories of a hillside privy belonging to her Great-aunt Lil and Great-uncle Bert Smith at Stoney Head, near Shipton Gorge. 'Their privy was not as prestigious as that of Auntie Lily in the New Forest, who had a two-holer, one big hole and one little hole side by side. But theirs was much more picturesque, reached by way of a tortuous and precipitous pathway up the steep hillside behind their cottage, from where it afforded a view of the traffic making its way along the busy A35.'

One evening in 1962, seven-year-old Christine convinced her mother that she needed to go one more time before leaving for home. 'It was winter time and already dark and frosty, pitch black in fact and without any street lights. Uncle Bert lit a hurricane lamp and led the procession up the sheer, winding path to the privy. It was like climbing the north face of the Eiger. I remember my mother wanted to come in with me but leave the door ajar, as she suffered from claustrophobia. There was no way I was going to sit with the door open so that every passing lorry driver could look up and see me silhouetted against the light of the hurricane lamp with my little pink nick-nacks round my knees, so I persuaded her that I would be fine on my own.'

The cottage was later demolished as part of a road improvement scheme, but 'their privy, further back and perched on the rocky hill slope, remained for many years, a forlorn landmark probably mistaken by passing tourists for a rustic shepherd's hut. Sadly, in the fullness of time it has mouldered away into the Dorset landscape.'

Mrs Doris Boswell-Smith, aged 86, was born and bred in Broadwindsor, seven miles inland from Bridport and West Bay. She, too, had to leave for the privy in good time. 'We lived in a cottage which was situated under a hill-field known then as Courtdown, and our garden followed the down slope of the hill behind. Our loo was in a corner at the top of the garden, a good 80 to 100 feet uphill from the back door of the cottage. My family always called the old loo "up garden" and I recall many times as a child – and indeed in my teens – when it became necessary to go "up garden". This meant walking up a path by a hedge which separated us from the uphill sloping field. My mother would light a candle in a candlestick and, with hands around the flame to save

It paid to get on with the neighbours if your privies were side by side. Friends could exchange the day's gossip through the wall while attending to other matters. This pair at Worth Matravers are made of Purbeck stone.

it blowing out, I plodded up the garden – complete with coat, hat and wellies if it was a wet night. When I was very young I had to be accompanied by an older sister who carried the matches, just in case!

'There was no door on the privy. It was an extension of a cow shed situated the other side of the hedge, with an earth floor on which was built a kind of box seat, about five feet long and two feet high, with two convenient holes in the top where two people could sit comfortably. There were no buckets. I can remember my mother and father putting soil and ashes down the holes.

'I well remember cutting squares of newspaper and making a hole with a skewer in the corner of each square, through which a piece of string was threaded and the whole hung on a nail in the wall of the privy.

This single seater is in Tolpuddle, back to back with another. Pigs occupied one half of the privy building – their sty can be seen on the right.

'The retired farmer who lived next door kept just one Jersey cow. In the winter it would be brought in from the weather to its stall adjoining our privy, and sitting there one listened to it chewing the cud and lowing – a wonderful childhood memory of our "up garden".

'I visited my old house recently and found it very up-to-date, with indoor plumbing, heating etc, but the happy memories of our "up garden" holes in the wood came flooding back. They remain with me and always will.'

[1 0]

MISCELLANEOUS MIXENS

One of the county's more exotic and little-known claims to fame is that a particular Dorset dunny was often used by a former Prime Minister of Australia and his First Lady. It arose from their friendship with PC Bill Edrich, who discovered that a policeman's lot is not a happy one while stationed as a country bobby at Loders, near Bridport, from 1949 to 1958.

Loders Police Station occupied a semi-detached farm cottage and conditions for Bill and his family were far below those in most police houses. Their water came from a cast iron pump in the back garden, shared by both cottages, which habitually froze in the winter. 'Indoors, there was no piped water, drain outlets or toilet facilities,' says Bill. 'We washed in a tin bowl on the kitchen table and bathed in the living room, in an old galvanised bath in front of the coal range – our only source of heat.

'There was a small privy in the garden with gaps at top and bottom of the door for ventilation. It contained a large iron bucket under a flat wood shelf with a generous oval hole in the middle. Fortunately, I had a fair sized garden in which to bury the obvious residue.

'Our children were still quite small and on one occasion my wife and I were at the back of the house, talking to our elderly next-door neighbour, when she suddenly pointed behind us and shrieked, "Missus, the baby!" We turned to see our daughter's little head and two feet just visible at the top of the hole in the privy seat. You can imagine where the rest of her was, and what it was like.'

While serving in submarines prior to joining the police, Bill had made friends with a family in Australia, which led to him offering hospitality to their daughter's boyfriend when he came

Many Dorset privies seem built to last for generations, with really thick stone walls. This one at Shipton Gorge stands back to back with its neighbour. The V-notches above the door were often the only ventilation, which explains why many people preferred to sit with the door open.

to study at Oxford. The young man's name was Bill Hawke and on several occasions he and his future bride stayed at Loders, braving the privy and other primitive facilities which Hazel Hawke later described so graphically in her autobiography, *My Own Life*. Bill Hawke went on to become his country's Prime Minister but he and Hazel have always remained friends with Bill Edrich, who retired with the rank of detective inspector.

Plenty of other fascinating tales came my way as I pursued privies the length and breadth of Dorset. John Loader wrote from Wimborne where, he explained, mains drainage was not available until the early 1960s. 'Apparently, our town council had rejected it many years before on the grounds of cost. It was said

It's not every day they slap a preservation order on a 'gents'. This unusual grade II listed public convenience (circa 1905) in Holdenhurst Road, Bournemouth, is described in the listing as 'a witty and carefully detailed solution to a utilitarian structure'. Various suggestions as to its original function include a tram controller's office – and a radio station!

that we never had any epidemic in Wimborne as a result of our earth closets.

'Collection was made once a week. The early morning collections were carried out by two well-known local characters, Tom Kerley and Dave Mabey, suitably attired in old clothes and hats, who rode on a large cylindrical tank. It was horse-drawn and formed a T shape, with shafts at the front. There was a wheel at each end of the cylinder and a hinged hatch in the middle at the top, with seats for the driver and his mate just in front of that.

'Stories abound regarding Tom and Dave, the most famous concerning an incident when they were driving along with a full load on their way to the dumping ground out of town by the river. Dave's sandwiches were in his bag by the open hatch. Alas, they fell out of the bag, straight through the hatch. Dave quickly grabbed them as they sank and remarked to Tom, "It's a good job they're wrapped, otherwise I wouldn't have had any breakfast."

'Often the houses had no rear access, which meant the collector had to go through the house from front door to back yard, tip the privy contents into his bucket and carry it back through the house to the tank out in the road. Many shops with living accommodation above also had no back entrance. Their doors were left unlocked so access could be gained to the back yard or garden loo.

'I recall what I was once told by John Gillingham, an elderly gentleman whose house had no back entrance. Tom Kerley had to enter his unlocked front door and walk along the passage inside the house to the back garden for the collection, passing the door to John's front room and his always open kitchen door. John, who took his breakfast early in the morning, told me, "I do like old Tom but I wish he would stop to yarn on his way in and not on his way back".'

George Samways still lives in Portesham, as he did when a child in a family of six served by a bucket privy. It was in the garden, against the boundary wall, backing onto the river and road, and his father used to rise at six in the morning to empty the bucket – into the river. The most popular time for other villagers was Saturday night.

One of George's earliest traumas came while he was sitting on its wooden seat, minding his own business. 'Suddenly a terrible scream came from beneath me, under the seat. I leapt off and shot out of the door, straight into a wall which cut my head open. You try running with your trousers round your ankles! It turned out that father had set a trap there for a rat that had been eating out of our bucket. Next door's terrier dog finished off the rat.

'In 1944, when the Americans were preparing for D-Day, one of their lorries ran away, crossed road and river and demolished our privy. Fortunately it was not in use at the time. We had to share with next door for a while. Finally it was rebuilt but then we had to wait for a new bucket to be found – they were scarce due to the war.

'The new privy had very little use before it too was knocked down, again by a lorry. It was rebuilt at a cost of £15. I still have the bill and a scar above my eye to remind me of our privies.'

An encounter with a rat is reported also by Mrs Roberts of Christchurch, who says she'll never forget the loo in their back yard. 'I was nine years old then, and opened the door to find a rat chewing up newspaper on the floor. Luckily, our fox terrier dog who used to follow me everywhere was behind me and in a flash she grabbed the rat. That was its very swift end. Although I cried my eyes out for the poor animal and wished it could have escaped, after that I always made sure my dog went in there first before I dared to shut myself in.

'Incidentally, our loo was also the best hiding place to escape from Mum and Dad when you were in trouble for misbehaving.'

Mischievous boys obtained endless entertainment from the rear emptying hatches of some bucket privies. Elsie Barter from Dorchester remembers her late husband telling of how youngsters in Cheselbourne used to plague an elderly deaf lady by opening the hatch while she was in residence and tickling her behind with feathers, making her scream and rush out in fright.

From David Burdett comes the tale of a former Commodore

of Weymouth Yacht Club who, as a schoolboy, would lift the
hatches and use a bunch of nettles to swish the bare backsides of
the sitting girls at a neighbouring school. 'Such pranks must
have happened often,' says David. 'In similar circumstances, I
once prodded my elder brother's genitalia with a stick. He
burst out of the privy, chased me and gave me a hiding which I
have never forgotten.'

A like story is told by John Whines of Morcombelake, who for
part of the war lived in Cerne Abbas. 'Many of the houses there
had privies at the bottom of their gardens, overlooking the River
Cerne,' he says. 'The local boys had great fun wading in the river
with bunches of stinging nettles when someone was sitting on the
privy.

'I moved down river to Godmanstone soon afterwards and we

A common design for Dorset privies is pictured here at Corfe Castle. Many
small homes had no gardens but did have vegetable plots elsewhere in the village, to which the buckets of manure were carried through the streets, usually
after dark (and very carefully).

had several cold winters, when the ground was too hard to dig to empty the privies. People would wait until after dark, then make their way across fields to empty their buckets in the Cerne.'

Marjorie Dawkins was 92 when she told me her memories of the one-hole privy at the end of the garden. 'If the weather was cold or wet, we had to wear a mac, hat and gloves. Some women did not like to be seen going to the toilet, so it was built behind the coal stack and women would take a coal bucket with them and collect coal afterwards. For the bedrooms we had china chamber pots. My mother had one with a large eye looking upwards from the bottom, with the words "What I see I never tell".'

When Mr H. Dyett met his future wife and went to her home in Hamworthy in 1952, he was shocked to find they still had a bucket privy. 'The first thing I did when we were married was to put a battery light in there as up to then the family had used a candle. The bucket was emptied twice a week by a man we called Ernie. He always had a sweet ready for our young son, so my wife had to watch out for him coming and run out, take the sweet off the boy and give him one of ours instead. One morning Ernie arrived early and caught my wife on the loo with the door open because of the odour. "That's all right, missus, I'll wait," he said. And he did.'

David Burdett recalls a 90-year-old retired headteacher telling of the time he was posted to a tiny village school in a remote part of Dorset. One day, a boy asked to 'be excused'. When he failed to return after a reasonable period, the teacher found him outside the playground privy, crying with shame as he had filled his pants. Why hadn't he used the privy? Because, sobbed the youngster, he hadn't managed to undo his collar in time.

A local colleague later told the mystified teacher that some years earlier a villager had been found dead in his privy, having suffered some form of fit or seizure. The doctor commented that he might not have died had someone been on hand to loosen his collar, a remark which soon spread round the village, and from that day no man or boy would dare attempt a bowel movement without first undoing the stud-fixed collar of his shirt.

Ben Gruitt of Bradpole, near Bridport, was born in 1923 and says, 'My parents rented a cottage in Morcombelake where the privy was a sort of stone shed down a long path into the orchard. I well remember the wood seat had two holes, one for adults and one for the children. The door would not stay shut and if one heard anyone else coming one had to put a foot against it to keep them out.

'There was a stump of wood coming out of the wall on which Mother would put a candle, and there was a pile of newspapers, mostly the *Daily Sketch* (only the "gentry" used toilet paper in those days). It was a communal privy and Mum always complained that the next door neighbour sat there and read, using up the candle.

'During the winter one had to light a lantern should one have to do a big job at night. Otherwise, it was behind the hedge.

'As children we often escaped to the bog and when time was up Dad used to blow a whistle. He always emptied the contents on his leave from the Navy. In those days, there was a huge trench in the garden over which they grew runner beans. Tomato plants also sprouted up and grew everywhere.

'During the first years of my marriage, we lived in a cottage in Middle Street, Bradpole, where the toilet was a bucket in a shed in the yard, emptied once a week by a man with a tanker who charged us two shillings and sixpence. Towards the end of the

A recently retired couple who moved from 'up north' intend to renovate their newly acquired privy at Hilton and perhaps thatch it to match the cottage.

week, we had to take stock of how full the bucket was, and finally we had to buy a second bucket to finish the week. Fortunately, I worked at a local school, so often I would wait until getting to work before doing the daily duty.'

Mrs Cherry Smeeth from Wimborne Minster remembers how thankful her family were to move, in 1947, to a new house in Churchill Road which had proper flushing toilets connected to the sewer. 'All the houses in the adjoining road, Hardy Crescent, still had bucket loos and on certain days there was a loud clanking noise as the lorries and men arrived to empty the buckets. There were great celebrations when the new sewer works were completed, and children thought it was enormous fun to pull the chain.'

[11]

THE FINAL FLUSH

Most people who experienced the delights of a back garden privy did so from necessity rather than from choice, and great was the rejoicing when their first WC was installed. But a size-able minority genuinely mourned the passing of the privy. Even today it is possible to find those who prefer to recycle their waste directly into the earth, believing the old way was the best.

Such a couple are Sally and Andy, who discovered an ancient, half-derelict cottage in a remote part of the Marshwood Vale, north of Bridport, and set about restoring it by the sweat of their own brows. In the cottage they found what their predecessors had used – a bucket surrounded by a portable wooden frame. Sally, who is 23, told me, 'We decided to continue using this system, as we like it.'

By the dexterous use of a funnel and a demijohn, she and her 30-year-old partner collect their urine separately, to be added to compost heaps in the garden. It has long been accepted that this adds valuable nutrients to the earth. As for solids, Sally explained: 'We have a wooden toilet seat over the bucket and we add wood shavings, earth or ashes, which we have found makes things much more acceptable. We empty the contents into pits at the end of the garden. The material composts down, then we use it in tree planting and on vegetable and flower beds.

'Most of the time we keep the bucket toilet outside in a large stone shed, sharing its space with fuel and garden tools. Fresh air and sunlight come in through the open doors and it is a pleasant place to sit. We move the bucket with its wooden framework back into the cottage during particularly cold weather.'

There must be others in secret parts of Dorset, as there are throughout the country, who prefer to observe the age-old

93

Too many people tear down their redundant outhouses when they could be preserved as attractive garden sheds and conversation pieces, like this one at Verwood.

customs. Sally outlined their philosophy: 'We enjoy the responsibility of dealing with our own waste, knowing that we are not over-using water or polluting the environment.' Good luck to them, say I.

Many people reading this book may be astonished at what their parents or grandparents took for granted in the not too distant past. In writing it, one of my motives is to pay tribute to the way in which they dealt with a basic and extremely pressing dilemma. Anyone who may be inclined to sneer dismissively should ponder how they would cope if they, too, were to be deprived of the luxuries of running water and mains sewerage.

And here's a thought for those who smugly believe our modern methods to be superior, more hygienic than those old fresh-air privies. It seems that every time you flush your oh-so-sophisticated WC, all that splashing water throws up an invisible cloud of nasty bacteria which float from the toilet bowl to contaminate everything within at least a six feet radius, including you. Where do you keep *your* toothbrush and face flannel?

Closing the lid doesn't help, in fact it can turn the WC into a powerful aerosol, forcing the germ-ridden spray out under pressure.

In 1992, a firm that makes paper towels performed some simple experiments to see what came out of a WC in normal usage. They placed a red dye just above the water line, set the nearby hand dryers running continuously and watched what happened. Within a few hours, the dye was found to have entered the dryers through their intakes and was being dispersed over users' hands. It could only have come from inside the bowl, mixed with who knows what unpleasantness.

A brick-built former privy and wash-house formed the basis of this little retreat in a Tolpuddle garden. The original door is at the back, and tools are stored in the privy.

Three years later, in 1995, *Reader's Digest* reported on the work of a microbiologist who knows all about bacteria in loos. It's his specialism. He also uses dye, blue this time, to confirm that the contaminated spray is thrown into the room whenever we flush. And there's more. He says that if you try to beat the bugs and flush with a close-fitting lid, the pressurised cloud can lurk within for up to *four hours*, just waiting to rush out and spray the next unfortunate who lifts the lid. Now there's a happy thought.

Many old folk couldn't see the sense of doing away with the outdoor privy and bringing unpleasant matters indoors. Perhaps they had the right idea after all.

To celebrate the long-awaited installation of flush toilets in their Dorset village, Tom's grandson invited him round to a family barbecue. As the old chap sat gazing down the path, the boy asked, 'Wossup, granfer? You ent hardly touched your food.' Tom shook his head. 'I were just thinkin' how it don't make sense. All my life I've had a privy in the yard and ate me grub in the house. Now we eats in the yard and goes indoors to shit. And they calls it progress.'